Project MIND
Math Is Not Difficult

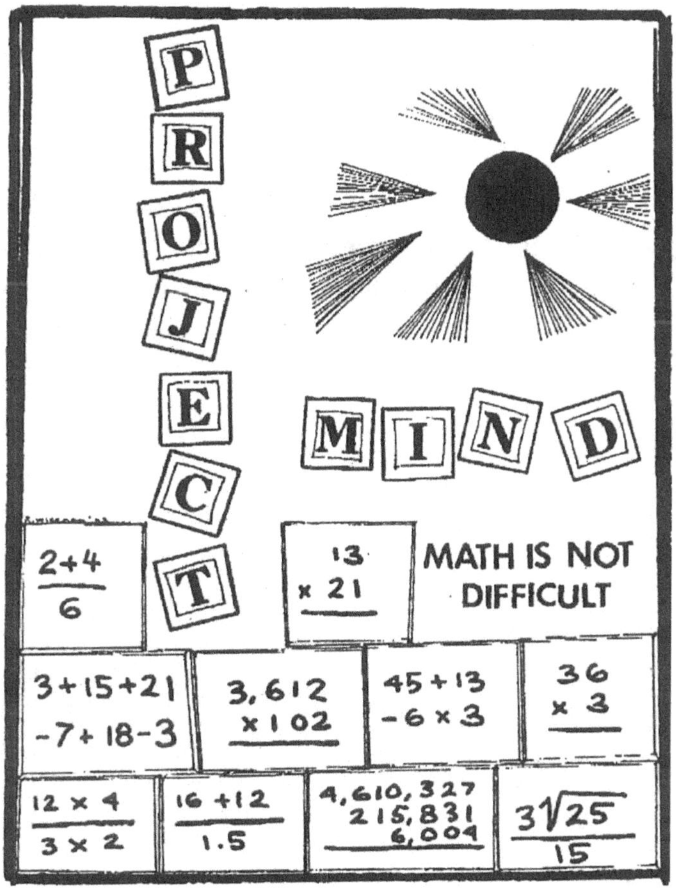

Pre-Kindergarten and Kindergarten
Mental Math Flash Cards

Hui Fang Huang "Angie" Su, Ed.D.

Project MIND, Inc.

The Mental Math Game

The students form two teams and come up to the bells two at a time. Upon looking at the math problem on a yellow card, they solve the problem mentally as fast as they can, usually within three seconds. The winner continues on while the loser moves to the end of his line. To be an intermediate champion, one must respond to three problems in a row correctly. After four intermediate champions are picked (depending on the size of your group, you must make sure that each student had a t least three chances), they are then entered into the second level of competitions with the green cards (more difficult problems.) To be a runner up for the grand champion title, competitors must also respond correctly to three problems in a row. Two runner-ups for the advanced level cards (red) are picked. They now compete for the title. The first person to respond to three red card problems in a row correctly is the grand champion.

Variations:

- The students compete in four areas: Mentally solve math problems with cards (visual aids), mentally solve math problems without cards, word problems, and equations (a string of problems to solve as the reader reads them.)
- The game can be played with your own class, another class, your grade level, or with other grade levels (fourth grade competing against fifth grade, third grade competing against fourth grade, etc.)
- If you have an advanced group, make sure that they use the cards for the next grade.
- Decimals and fractions can be added for third through fifth grade.

Pre-Kindergarten/Kindergarten:

- Level 1 – Yellow Cards: Number identification and shape identification
- Level 2 – Green Cards: Number identification (up to 100), identify the missing number, and adding and subtracting up to 5.
- Level 3 – Red Cards: number sequencing, and adding and subtracting up to 10
- Equations: strings of numbers which add and subtract up to 10
- Word problems: Simple one step, how many items? Adding or subtracting up to 10

First Grade:

- Level 1 – Yellow Cards: Adding and subtracting numbers up to 10
- Level 2 – Green Cards: Adding and subtracting two-digit numbers and adding three digit numbers
- Level 3 – Red Cards: Adding and subtracting three-digit numbers

Second Grade:

- Level 1 – Yellow Cards: Adding and subtracting two-digit numbers

- Level 2 – Green Cards: Adding and subtracting two-digit numbers with carrying and borrowing, and multiplication and division facts
- Level 3 – Red Cards: Adding and subtracting three-digit numbers with carrying and borrowing; two-digit multiplication

Third Grade:

- Level 1 – Yellow Cards: Adding and subtracting two-digit numbers with carrying and borrowing; single digit multiplication and division
- Level 2 – Green Cards: Adding and subtracting three-digit numbers with carrying and borrowing, and two-digit multiplication and division
- Level 3 – Red Cards: Adding and subtracting four-digit numbers with carrying and borrowing; three-digit multiplication and division

Fourth Grade:

- Level 1 – Yellow Cards: Adding, subtracting, multiplying, and dividing fourth grade level problem
- Level 2 – Green Cards: Adding, subtracting, multiplying, and dividing fourth grade level problems that are harder than Level 1
- Level 3 – Red Cards: Adding, subtracting, multiplying, and dividing multi-digit fifth grade level problems

Fifth Grade:

- Level 1 – Yellow Cards: Adding, subtracting, multiplying, and dividing fifth grade level problem
- Level 2 – Green Cards: Adding, subtracting, multiplying, and dividing fifth grade level problems that are harder than Level 1
- Level 3 – Red Cards: Adding and subtracting six digit numbers with carrying and borrowing, and multiplying and dividing multi-digit problems

1

2

3

Project MIND

4

Project MIND

5

6

7

8

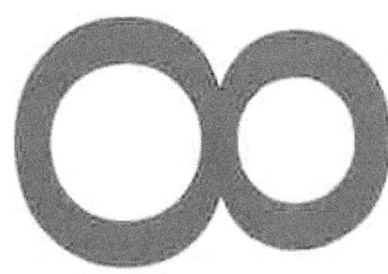

Project MIND

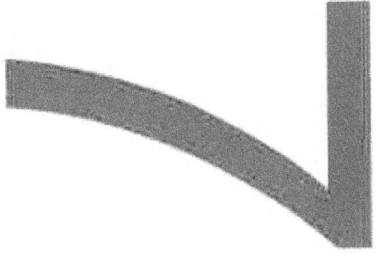

9

10

11

12

Project MIND

13

14

15

Project MIND

16

Project MIND

17

18

19

20

Project MIND

21

Project MIND

22

Project MIND

23

24

25

26

Project MIND

27

28

Project MIND

29

30

31

32

Project MIND

33

34

35

Project MIND

36

Project MIND

37

38

39

40

41

42

43

44

45

Project MIND

46

Project MIND

47

48

49

50

Project MIND

51

52

53

54

Project MIND

55

56

Project MIND

57

58

59

Project MIND

60

Project MIND

61

62

63

64

Project MIND

65

66

67

68

Project MIND

69

Project **MIND**

70

71

Project MIND

72

Project MIND

73

74

75

76

Project MIND

77

78

Project MIND

79

80

Project MIND

81

82

83

84

85

86

87

88

89

90

Project MIND

91

92

93

94

95

96

97

Project MIND

98

Project MIND

99

100

Project MIND

15, ______ ,17

______ ,18,19

16

Project MIND
Kindergarten – Green

17

Project MIND
Kindergarten – Green

_____, 8, 9

7, 8, _____

7

Project MIND
Kindergarten – Green

9

Project MIND
Kindergarten – Green

2, ______, 4

______, 3, 4

3

Project MIND
Kindergarten – Green

2

Project MIND
Kindergarten – Green

______,17,18

13,14,______

16

Project MIND
Kindergarten – Green

15

Project MIND
Kindergarten – Green

12,13,______

6, 7, ______

14

Project MIND
Kindergarten – Green

8

Project MIND
Kindergarten – Green

5, ______, 7

______, 2, 3

6

Project MIND
Kindergarten – Green

1

Project MIND
Kindergarten – Green

14, ________ ,16

18, ________ ,20

15

Project MIND
Kindergarten – Green

19

Project MIND
Kindergarten – Green

7, ____, 9

1, 2, ____

8

Project MIND
Kindergarten – Green

3

Project MIND
Kindergarten – Green

8, 9, ______

4, ______, 6

10

Project MIND
Kindergarten – Green

5

Project MIND
Kindergarten - Green

$$\begin{array}{r} 1 \\ + \quad 1 \\ \hline \end{array}$$

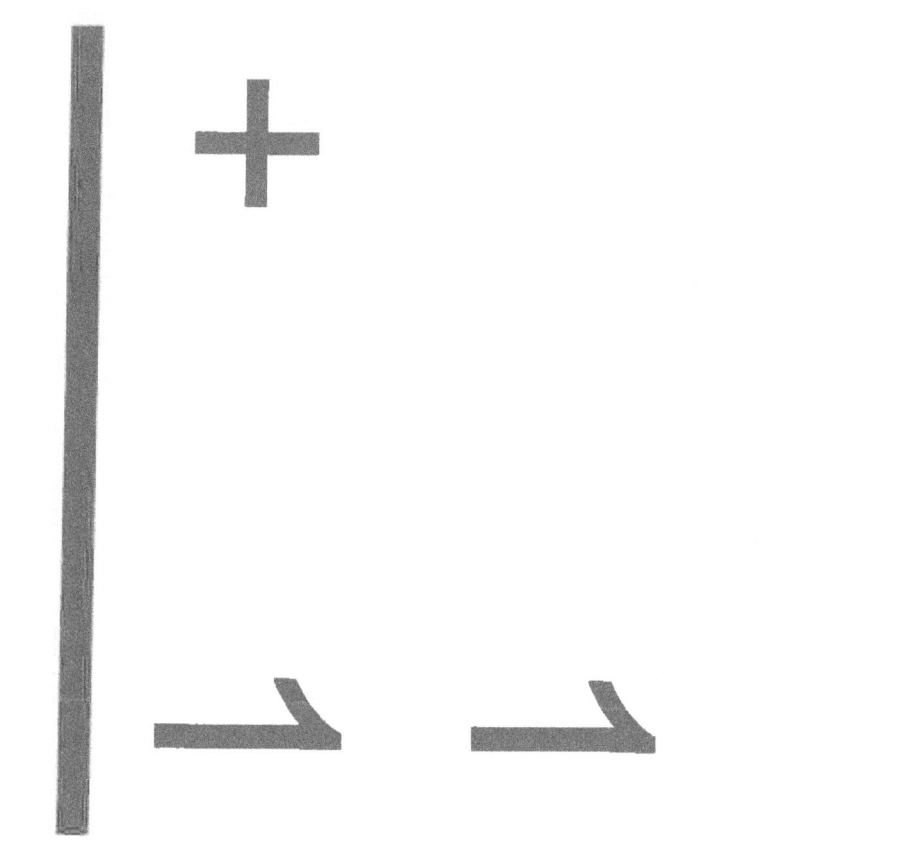

$$\begin{array}{r} 1 \\ + \quad 2 \\ \hline \end{array}$$

$$\begin{array}{r} 1 \\ +\ 2 \\ \hline \boxed{3} \end{array}$$

Project MIND
Kindergarten - Red

$$\begin{array}{r} 1 \\ +\ 1 \\ \hline 2 \end{array}$$

Project MIND
Kindergarten - Red

$$\begin{array}{r} 2 \\ +\quad 2 \\ \hline \end{array}$$

$$\begin{array}{r} 2 \\ +\quad 1 \\ \hline \end{array}$$

$$\begin{array}{r} 2 \\ +\ 1 \\ \hline 3 \end{array}$$

Project MIND
Kindergarten - Red

$$\begin{array}{r} 2 \\ +\ 2 \\ \hline 4 \end{array}$$

Project MIND
Kindergarten - Red

$$\begin{array}{r} 3 \\ +\quad 2 \\ \hline \end{array}$$

$$\begin{array}{r} 3 \\ +\quad 1 \\ \hline \end{array}$$

$$\begin{array}{r} 3 \\ +\quad 1 \\ \hline \boxed{4} \end{array}$$

Project MIND
Kindergarten - Red

$$\begin{array}{r} 3 \\ +\quad 2 \\ \hline 5 \end{array}$$

Project MIND
Kindergarten - Red

$$\begin{array}{r} 5 \\ + \quad 1 \\ \hline \end{array}$$

$$\begin{array}{r} 4 \\ + \quad 1 \\ \hline \end{array}$$

$$\begin{array}{r} 4 \\ + \quad 1 \\ \hline \boxed{5} \end{array}$$

Project MIND
Kindergarten - Red

$$\begin{array}{r} 5 \\ + \quad 1 \\ \hline 6 \end{array}$$

Project MIND
Kindergarten - Red

$$\begin{array}{r} 6 \\ + \quad 4 \\ \hline \end{array}$$

$$\begin{array}{r} 5 \\ + \quad 2 \\ \hline \end{array}$$

$$
\begin{array}{r} 5 \\ +\ 2 \\ \hline \boxed{7} \end{array}
$$

Project MIND
Kindergarten - Red

$$
\begin{array}{r} 6 \\ +\ 4 \\ \hline 10 \end{array}
$$

Project MIND
Kindergarten - Red

$$\begin{array}{r} 2 \\ -\ 1 \\ \hline \end{array}$$

$$\begin{array}{r} 1 \\ -\ 1 \\ \hline \end{array}$$

$$\begin{array}{r} 1 \\ -\ 1 \\ \hline 0 \end{array}$$

Project MIND
Kindergarten - Red

$$\begin{array}{r} 2 \\ -\ 1 \\ \hline 1 \end{array}$$

Project MIND
Kindergarten - Red

$$\begin{array}{r} 3 \\ -\ 2 \\ \hline \end{array}$$

$$\begin{array}{r} 3 \\ -\ 1 \\ \hline \end{array}$$

$$\begin{array}{r} 3 \\ -\ 1 \\ \hline 2 \end{array}$$

$$\begin{array}{r} 3 \\ -\ 2 \\ \hline 1 \end{array}$$

Project MIND
Kindergarten - Red

$$\begin{array}{r} 4 \\ -\ 2 \\ \hline \end{array}$$

$$\begin{array}{r} 4 \\ -\ 1 \\ \hline \end{array}$$

$$\begin{array}{r} 4 \\ -\ 1 \\ \hline 3 \end{array}$$

Project MIND
Kindergarten - Red

$$\begin{array}{r} 4 \\ -\ 2 \\ \hline 2 \end{array}$$

Project MIND
Kindergarten - Red

$$\begin{array}{r} 5 \\ -\ 2 \\ \hline \end{array}$$

$$\begin{array}{r} 4 \\ -\ 3 \\ \hline \end{array}$$

$$\begin{array}{r} 4 \\ -\ 3 \\ \hline 1 \end{array}$$

Project MIND
Kindergarten - Red

$$\begin{array}{r} 5 \\ -\ 2 \\ \hline 3 \end{array}$$

Project MIND
Kindergarten - Red

$$\begin{array}{r} 6 \\ -\ 4 \\ \hline \end{array}$$

$$\begin{array}{r} 5 \\ -\ 5 \\ \hline \end{array}$$

$$\begin{array}{r} 5 \\ -\ 5 \\ \hline 0 \end{array}$$

Project MIND
Kindergarten - Red

$$\begin{array}{r} 6 \\ -\ 4 \\ \hline 2 \end{array}$$

Project MIND
Kindergarten - Red

www.ingramcontent.com/pod-product-compliance
Lightning Source LLC
Chambersburg PA
CBHW081832170526
45167CB00007B/2793
* 9 7 8 1 5 0 3 2 6 9 4 4 6 *